SpringerBriefs in Fire

Series editor

James A. Milke, College Park, MD, USA

For further volumes:
http://www.springer.com/series/10476

Anca Taciuc · Anne S. Dederichs

Determining Self-Preservation Capability in Pre-School Children

Anca Taciuc
Anne S. Dederichs
Technical University of Denmark
Kongens Lyngby
Denmark

ISSN 2193-6595 ISSN 2193-6609 (electronic)
ISBN 978-1-4939-1079-3 ISBN 978-1-4939-1080-9 (eBook)
DOI 10.1007/978-1-4939-1080-9
Springer New York Heidelberg Dordrecht London

Library of Congress Control Number: 2014939062

Printed on acid-free paper

Springer is part of Springer Science+Business Media (www.springer.com)

Foreword

The self-preservation capability of preschool children is an uncommon topic, there being little data on evacuation characteristics of children. Self-preservation capability is defined as the ability of a client to evacuate a location [1], in this case a day-care occupancy, without direct intervention by a staff member, having as example: persons who are not able to use the stairs or cannot follow directions to go outside of a facility. As examples for intervention from staff members there are: carrying, guiding by direct hand-holding, or continued bodily contact with the client.

The purpose of the current project is to determine an age-limit for preschool children at which they may be considered being capable of self-preservation. Furthermore, the goal is to gain a better understanding of child behavior and also to have an overview of the evacuation procedures in case of fire in the respective day-care centers. To find these answers, an international survey among teachers from day-care centers and experts in child development was carried out.

Teachers and experts have a different opinion with respect to preschool children preservation. Taking into consideration data received from teachers and experts from all countries, the age at which a majority of children are considered capable of self-preservation is between 30 and 36 months. At this age, a majority of children are considered able to understand and follow simple instructions, walk on horizontal surface without physical support, and walk down stairs.

Keywords Life safety code®, Children, Evacuation capability, Self-preservation

Preface

In the 1994 edition of NFPA 101, Life Safety Code®, the requirements for Day-Care Centers were completely rewritten by the Technical Committee on Assembly and Educational Occupancies to bring the requirements more in-line with the current (at the time) functioning of day-care centers and to include adult day-care centers into the definition of a day-care center. Part of the revision work included restructuring requirements based on capability of self-preservation. For children, the Task Group that performed the initial investigative work for these revisions interviewed multiple day-care center owners and early childhood experts to determine at what age toddlers would be considered incapable of self-preservation. The questions answered by the early childhood experts were as follows:

1. At what age will a child take instruction from staff and follow those instructions without having to be carried or using hand-holding techniques, following in a line the staff member to the outside of the building?
2. At what age will most children be able to walk up or down stairs without having to be carried or have to drop to their knees in order to climb up or down stairs?

The bulk of the answers received by the Task Group was 24 months. Hence, the Technical Committee defined children younger than 24 months as incapable of self-preservation and thus more stringent requirements in the Code apply. It has been noted that the International Fire Code uses 30 months for their cut-off for "self-preservation," even though they do not use that specific term. It means the difference between being classified as "I" (Institutional), which has more stringent requirements versus "E" (Educational). More information was needed on the topic, so a project was undertaken that included a review of the literature on the topic and a survey of early childhood experts and teachers.

Acknowledgments

The Fire Protection Research Foundation expresses gratitude to report authors Anca Taciuc and Anne S. Dederichs of the Technical University of Denmark. The Research Foundation appreciates the guidance provided by the Project Technical Panel: Rita Fahy, Cathy Stashak, Alex Szachnowicz, and Ron Cote. Special thanks are expressed to the National Fire Protection Association (NFPA) for funding this project.

Acknowledgments

Contents

About the Fire Protection Research Foundation

The Fire Protection Research Foundation plans, manages, and communicates research on a broad range of fire safety issues in collaboration with scientists and laboratories around the world. The Foundation is an affiliate of NFPA.

About the National Fire Protection Association (NFPA)

NFPA is a worldwide leader in fire, electrical, building, and life safety. The mission of the international nonprofit organization founded in 1896 is to reduce the worldwide burden of fire and other hazards on the quality of life by providing and advocating consensus codes and standards, research, training, and education. NFPA develops more than 300 codes and standards to minimize the possibility and effects of fire and other hazards. All NFPA codes and standards can be viewed at no cost at www.nfpa.org/freeaccess.

Chapter 1
Introduction

Between 2005 and 2009, fire departments from the U.S. reported an average of 590 fires annually in day-care centers [2], having an annual average of 8 civilian fire injuries and $4.5 million in direct property damage. The majority of fires occurred between 6:00 a.m. and 3:00 p.m. and the primary cause involved cooking equipment (64 %).

Fire incidents in educational properties happen every year. In 2004, there was a school fire in Kumbakonam, India where 87 children died and 27 were injured, because the thatched roof caught fire [3]. The firefighters and the rescue operators were impeded by the lack of access into the three story building, which had only one entrance and a single flight of stairs. In 2009, a daycare center fire occurred in Mexico where 47 young children lost their lives because of the flame spread from a nearby storehouse which didn't have any appropriate security measures [4].

Building codes' and fire regulations' primary aim is to prevent human losses. The objective of this current international project was to determine at what age children are considered to be capable of self-preservation and also to investigate the fire safety and evacuation measures for day-care centers. The study was conducted in USA, Denmark, Canada, Germany, France, Spain and Romania.

The following research questions form the basis of the study:

- At what age can the majority of children be expected to understand and follow simple instructions?
- At what age can the majority of children be expected to walk?
- At what age can the majority of children be expected to walk down the stairs?
- How do the fire safety installations, number of fire drills and ratio between staff and children differ between the different countries?

A literature review on the evacuation of young children, children psychology, national regulations for fire safety and the educational system in the countries where the questionnaire was distributed are presented in the next section. This is followed by a presentation of the method applied when developing the questionnaire. The results are presented and discussed and conclusions are drawn.

A. Taciuc and A. S. Dederichs, *Determining Self-Preservation Capability in Pre-School Children*, SpringerBriefs in Fire, DOI: 10.1007/978-1-4939-1080-9_1,

Chapter 2
Literature Review

Literature from different fields is relevant for the project. Besides self-preservation, evacuation procedures in day-care centers play a role for the overall process of evacuation, as well as children's psychology, providing information on the age range at which children can be expected to be able to carry out self-preservation.

Fire building regulations for day-care centers and educational system were studied for each country where the questionnaire was distributed to understand and discuss the final results of the questionnaire.

2.1 Evacuation Procedures of Young Children

Limited number of studies covers the topic of self-preservation of children in day-care centers and little is known on how their behavior affects the evacuation process and total evacuation time.

A study on fire safety and evacuation planning for pre-schools and day-care centers was carried out , where 70 % of the pre-schools and day-care centers are located in multi-story buildings and 30 % of cases the infants' rooms were located on the upper floor. The results come from a questionnaire and fire drills. The work has the following findings:

- More than one evacuation route is recommended.
- As well as every day usage of the routes.
- The adult-child ratio necessary for efficient evacuation was determined for each day-care center.
- One fire drill per month is required, however, 70 % of the day-care centers performed a maximum 7–8 fire drills per year.

One of the conclusions of the study is that familiarity with the system and procedures is the most important factor affecting the speed of evacuation.

Data was collected from two evacuation drills, which were carried out in the same school building with children between 4 and 10 years old. The pre-

A. Taciuc and A. S. Dederichs, *Determining Self-Preservation Capability in Pre-School Children*, SpringerBriefs in Fire, DOI: 10.1007/978-1-4939-1080-9_2,

Table 2.1 Travel speed for differentiated age groups

Age groups (years)	Average walking speed (m/s)	Average run speed (m/s)	Adult average walking speed (m/s)
0–2	0.60	1.14	1.2–1.3
3–6	0.84	2.23	

movement time is strongly affected by the actions and decisions of the teachers. For the age category 4–5 years, the pre-movement depends on the teacher actions (encouraged and prepared for the evacuation) and for age group 5–6 years and primary school children are dependent on teacher decisions (children were responding quickly and forming a row by themselves, but they stopped and waited for teacher signal to go out). The drill results suggested that walking speeds on stairs are age dependent and it is noted that 72 % (Drill 1) and 66 % (Drill 2) of pre-school children used the handrail during the fire drill and also 44 % (Drill 1) of children moved downstairs with one foot/step while the percentage is greater for Drill 2–67 %.

Another study [7] was performed at the Technical University of Denmark regarding the movement of pre-school children from 10 day-care centers from Lyngby, Denmark, on stairs and on horizontal planes, focusing on flow, densities and walking speeds. The report compared the results with the ones found in the current literature for adults. For the horizontal travel speed, results show that 78 % of the young children (0–2 years) have an average walking speed of 0.60 m/s and respectively 0.84 m/s for more than 66 % of the older children (3–6 years). The common values used for adult's average walking speed are 1.2–1.3 m/s. These values can be observed in Table 2.1.

For the stair movement results, data were only from older children since the younger children's rooms were located on the ground floor. The study involves three different spiral stairs: first is used by children every day, having an extra convenient handrail for children, second is not used regularly and has an inconvenient handrail and the third is a metallic external fire escape—never used, where the steps are see-through having a handrail not adapted for children. Although the three stairs have similar dimensions, there is a large difference in the average travel speed. Results can be observed in Table 2.2. The conclusion is that travel speed is directly dependent on the familiarity and design of the stairs and handrails.

Results of the study reveal that the flow of children through doors is higher than the reference data for adults and the children didn't have any problem passing through doors two at a time, even if the width of the door had only 0.6 m.

Various evacuation drills were performed in Lyngby Taarbæk Commune in Denmark in 10 day-care centers with children between 0–2 and 3–6 years. The work had the following findings:

- The level of warning sound was considered insufficient.
- 5 of 10 day-care centers used verbal warnings, leading to a warning time of 31–265 s.

Table 2.2 Results on Spiral Stairs

Stair	Width (m)	Slope (°)	Average travel speed (m/s)
Stair 1	0.80	33	0.58
Stair 2	0.87	33	0.38
Stair 3	0.91	30	0.13

- The average walking speed on a horizontal plane was found to be 0.63 m/s for younger children (5 % running and 95 % walking) and 1.40 m/s for the older children (40 % running and 60 % walking).

Furthermore, it was found that for the age group:

- 0–2 years: 22.2 % were carried by staff, 57.6 % received some physical help, 20.2 % received only a verbal command;
- 3–6 years: 1.8 % were carried by staff, 12.3 % got some physical help and 85.9 % received only a verbal command.
- The average adult-child ratio during the fire drills was 1:3.2 for the age group 0–2 years and 1:6.1 for age group 3–6 years.
- Children followed instructions without questioning them, but they seemed surprised to be going outside without putting on shoes and jackets as they were used to. Also, they appeared to be affected by the use of unfamiliar routes. Children had the tendency to follow the daily routine: putting on shoes before going outside, stopping in front of the door before exiting to zip up their jackets or stopping in front of the main door waiting on an adult. It was noticed that many of the day-care centers had doors with a handle that the children could not reach.
- Teachers typically started by instructing the children to evacuate and in some cases they took a roll call before going out of the room. They took a list with the names of the children and phone numbers outside. Some of them were not aware that they should close the windows and doors in order to minimize fire spread. The procedure is that once evacuated, children should meet outside with an adult for a roll call to see who is missing, but not all adults knew this and children were running outside on the playground.
- Another important aspect was the timing of the fire evacuation drill. It is known that people are more vulnerable when sleeping [8]. In Denmark, children under the age of two sleep in cribs, placed outside or in a special shed. Children over the age of two sleep on mattresses indoors. At one institution, the first drill was done in the morning and the second one in the afternoon right when children woke up or were still asleep, which caused confusion. Cribs had to be rolled away.

In 2004, performance-based fire safety requirements were introduced in Denmark. The prescriptive part of the Danish Building Regulations subdivides buildings into different categories of usage [9]. Day-care centers belong to usage category 6. The installation of an automatic fire alarm system, notifying the fire department and the staff, is required. This rule does not apply for buildings constructed before 2004.

Pre-movement time for pre-school institutions, normal walking speed, running speed, upward and downward movement on stairs, and flow through door openings for children of different ages were determined in a Russian study from 2012 [10]. Fire safety training of the staff and the evacuation procedure affected the pre-movement time. It was observed that after hearing the fire signal, many teachers went to the corridor to verify that an evacuation had to be conducted. In the same study, children remained in their places without taking any evacuation measures, after the indication for evacuation was given. Evacuation first started when teachers took the kids by hand and led them outside. The observations made on the pre-movement time show that preparation time depends on the season. During cold weather storing blankets outside can considerably reduce the pre-movement time. The proposed time of preparation in summer time is 0.6, 5 min in spring and autumn, 7.5 min in winter if using outdoor clothing and 1.1 min when using blankets.

Literature contains different values for walking speeds on horizontal planes and down stairs [7, 10]. According to [7], the average speed is 0.84 m/s for the 3–6 year age interval. The average for the three age groups 3–4, 4–5 and 5–7 years, the speed is 0.83 m/s [10], concluding that even if the children's age intervals are different, the average horizontal walking speeds are comparable. In the case of average speed for walking downstairs, the results are hard to compare because of different conditions: the usage of stairs, geometry, handrail and children's age (Table 2.3).

In some buildings housing pre-school institutions, the staircases are designed for adults and are not equipped with railing on both sides and this impedes children's movement [10].

One study found that the staff was poorly trained with respect to fire evacuation [11]. 100 % of the teachers from one school were not aware about the fire safety system at their institution. 69 % of the staff failed to follow the valid evacuation procedure.

Data was gathered during semi-unannounced evacuation drills in Copenhagen [12], where 127 children between 0 and 6 years old were involved. The conclusion of the article is that the evacuation drill was age dependent—older children were rushing out meanwhile the young ones seemed to be more confused. Features that caused a evacuation delay included:

- Children could not open doors;
- In many cases there was no automatic audio alarm system but the evacuation was initiated using verbal warning.

Table 2.3 Comparison between results from two different drills

Age Groups (years)	Average horizontal (m/s)	Average downstairs (m/s)	References
0–2	0.60	–	[7]
3–6	0.84	Depending on type of stairs 0.13, 0.38, 0.58	
3–4	0.77	–	[10]
4–5	0.85	0.66	
5–7	0.86	0.73	

Table 2.4 Children's skills

Skill	Examples	
Physical	Gross motor	Stand, walk, run
	Fine motor	Eat, dress, play, write
Language	Speaking, understanding others and communication	
Cognitive	Learning, understanding, problem-solving, remembering	
Social	Interacting with others, relationships with family, friends or teachers	

Table 2.5 Development milestones

Age (months)	Skill	Definition
12–18	Gross motor	Walks alone, walks up stairs with help
	Psychological	May show dependence on primary caregiver does not respond well to sharp discipline and does not respond to verbal persuasion and scolding
18–24	Gross motor	Runs stiffly, comes downstairs on bottom or abdomen
24–30	Gross motor	Walks up and down steps, both feet on each step, runs headlong
	Language	Speaks 50 or more words has vocabulary of 300 words

2.2 Child Development

Child behavior and development milestones are assessed to play a role in evacuation situations. The focus of the studies presented here are children between 12 and 36 months.

Child development could be defined as how a child becomes able to do more complicated tasks as he is getting older [13]. Stages in children development can be observed in Table 2.4.

Development milestones are a set of skills or tasks specific at some age that most of the children can do between an age intervals as shown in Table 2.5, [14].

2.3 Fire Building Regulation and Educational System

Building fire regulations are an important factor determining the fire safety of day-care centers. The primary objective of these regulations is to prescribe requirements related to design and maintenance of the fire safety performance. The focus will be on the usage category for day-care centers, fire safety installations for these institutions, staff instructions, and fire drills. The regulations of three different countries are discussed below:

2.3.1 Denmark

In Denmark, day-care centers are classified in usage Category 6 according to Building Regulation [9], that is defined as building sections for day time and some case night time where the people which are occupying the building need help to evacuate themselves to a place of safety. For this category, an automatic fire alarm system, that warns the staff only is mandatory, but this regulation applies only for buildings that are built after 2004. This is why many day-care centers in Denmark do not have any alarms, warning systems or smoke detectors. Additionally is not mandatory for the day-care institutions to perform evacuation exercises or fire drills [8, 9]. As for the educational system in Denmark, parents begin to enroll their children from the age of 6 months in day-care facilities.

In the year 2004, Denmark replaced prescriptive fire regulations with performance based codes—which gives building designers more freedom in terms of number of exits, fire system etc.

2.3.2 United States

The Rules for the Licensing of Child Care Facilities, provided by Maine Department of Health and Human Services—USA (which is the authority having jurisdiction in that state), specify that at least once a month fire evacuations drills must be conducted for all the children and adults present, using at least two means of exits and also that the fire drills should be recorded for further inspections. Also the National Fire Protection Association's Life Safety Code® (NFPA 101) sets requirements that should be followed by new and already existing day-care institutions. Depending on the construction type and the building height, sprinkler systems might be required.

Table 2.6 Adult-child ratio Canada

Adult	Children	Children's age (months)
At least 1	<6	>18
At least 2	≥6	
At least 1	<4	<18
At least 2	≥4	

Table 2.7 Adult-child ratio Romania

Adult	Children	Children's age (months)	Observations
1	4	<12	Maximum of 7 children in one class
1	5	12–24	Maximum 9 children in one class
1	6	24–36	Maximum 9 children in one class

2.3.3 Adult-Child Ratio in U.S./Canada/Romania

The adult-child ratio varies in different countries. An adult-child ratio depending on the type of day-care facility is defined in the United States:

- Family day-care homes: 1:6 and should be not more than two children incapable of self preservation,
- Group day-care homes: 2:12 children and there should not be more than three children incapable of self-preservation [1, 15].

In Canada, staff members from the day-care facility are to be instructed with his/her responsibilities in case of a fire before beginning work and a fire drill is conducted at least once in a month [16]. In addition, the staff numbers and group size that are required are shown in Table 2.6.

In Romania [17], children are divided under age consideration and the adult-child ratio is (Table 2.7).

Chapter 3
Method

In the current chapter the method used to develop and distribute the questionnaire is described. The questionnaire was distributed amongst people working at day-care centers as well as child development specialists from the following countries: USA, Canada, Denmark, Germany and Romania. The goal was to determine at what age children can be considered capable of self-preservation.

3.1 Questionnaire

In order to get answers to the research questions, a questionnaire with 13 questions was developed. The questionnaire was evaluated by a psychologist, specialized in children development, from Lund University, Sweden and the panel members of the NFPA's Fire Protection Research Foundation Project Technical Panel. Two sets of questionnaires were developed: one aiming at workers in day-care institutions—Appendix A and second for the psychologists—Appendix B.

The questionnaire was translated into German, Danish, French, Spanish, and Romanian.

3.2 Participants and Distribution

The invitation to participate to the web based questionnaire was through email contacts (Appendix C). The questionnaire was aimed at professionals working in day-care centers, having experience in working with kids on one hand and also for experts in child development, such as: psychologists or physical therapists, which have rather theoretical knowledge.

The questionnaire was distributed in the following countries: United States, Canada, Denmark, Germany, Spain and Romania; through social network groups for child-care, University teachers from the department of Psychology and email addresses from day-care centers' websites.

A. Taciuc and A. S. Dederichs, *Determining Self-Preservation Capability in Pre-School Children*, SpringerBriefs in Fire, DOI: 10.1007/978-1-4939-1080-9_3,

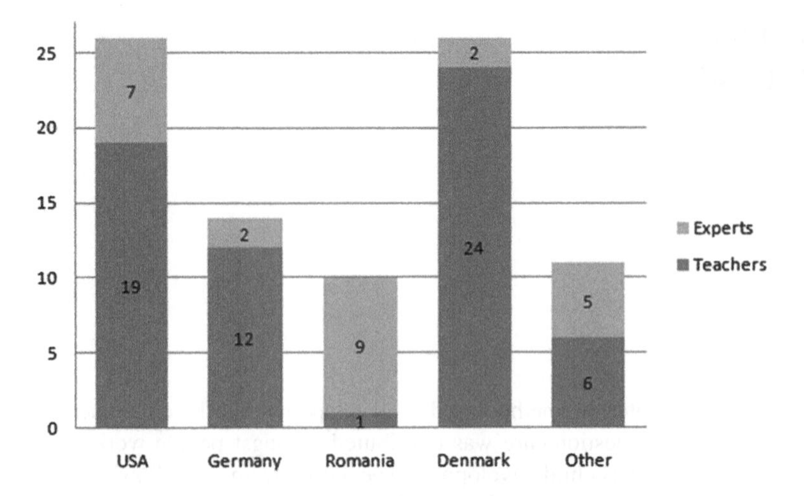

Fig. 3.1 Number of professionals that responded to the questionnaire

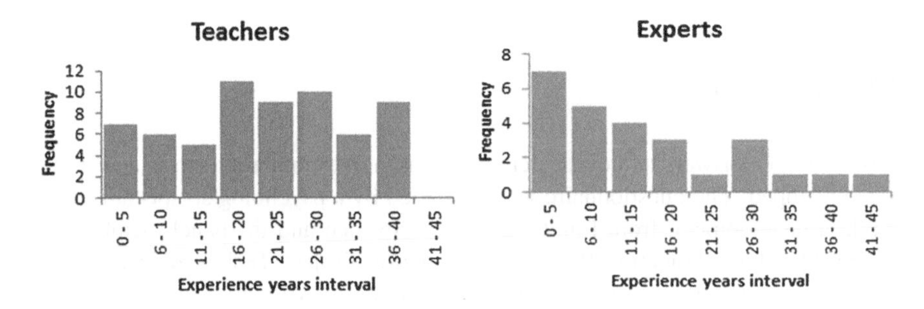

Fig. 3.2 Participants' experience

A total number of 87 participants answered the questionnaire. In the following, the term *teacher* refers to a professional working in a day-care center, having experience with children, and *expert* refers to a professional working in children development. Because of different professional experience, most of the responses given by teachers and by experts are different, that is why a term of "total" is added to take into consideration both opinions. Because of the unequal number of answers given by the teachers/experts or by country, also are mentioned the percentage and the number of participants who have answered. The term *mean* refers to the average value and term *mode* refers to the value which was most used.

Not all of the participants responded to all the questions. Categorizing them by their profession: 62 teachers and 25 experts participated in this study. 26 participants were working in USA, 14 in Germany, 26 in Denmark, 10 in Romania and 11 in different countries such as: Canada (4), Japan (1), UK (2), Australia (1) and Spain (3), see Fig. 3.1.

Figure 3.2 shows the participating teachers' and experts' experience with education and knowledge in growth and development of infants, toddlers and preschoolers.

Chapter 4
Results and Discussion

In this chapter, the results from the answers of the questionnaire are shown and discussed, comparing the answers given by the teachers and experts, the results from different countries and total results.

The results were classified by the profession of the participants and the country where they are working. The number of participants varied for the different countries. The displayed answers are restricted to countries that have more than 10 teacher participants (USA, Denmark, Germany) and more than 7 expert participants (USA, Romania).

As it can be seen in Fig. 4.1, the majority of the teachers believe that most of children can understand and follow simple instruction by the age of 36 months ("30–36 months"), meanwhile 56 % of the experts believe that they can by age of 30 months. Combining both opinions, it can be concluded that the majority of children can follow simple instructions by the age of 36 months.

In Fig. 4.2 it can be observed that over 50 % of teachers from USA and Germany believe that the majority of children are able to follow simple instructions by the age of 30 months. Similarly to the teachers' opinion the majority of experts from USA, Denmark and Germany answered by the age of 30 months see Fig. 4.3. Meanwhile the teachers from Denmark have answered ">36 months" and those from Germany "<24 months".

In Fig. 4.4 it can be observed that 69 % of teachers consider that the majority of children under 24 months are able to walk on a horizontal surface without any physical support, instead, 44 % of the experts consider by the age of 30 months. After the cumulative calculation, taking into consideration the teachers' and experts' opinion, the majority answer is "<24 months".

Hundred percentage of teachers from Germany believe that the majority of children are able to walk alone before the age of 24 months also smaller majorities of US and Danish teachers say before 24 months. The majority of experts from USA and Denmark consider that children are able to walk by the age of 30 months, see Figs. 4.5 and 4.6.

Teachers consider that by the age of 30 months (24–30 months) and experts by age of 36 months that children are capable of walking down stairs. The cumulative

A. Taciuc and A. S. Dederichs, *Determining Self-Preservation Capability in Pre-School Children*, SpringerBriefs in Fire, DOI: 10.1007/978-1-4939-1080-9_4,

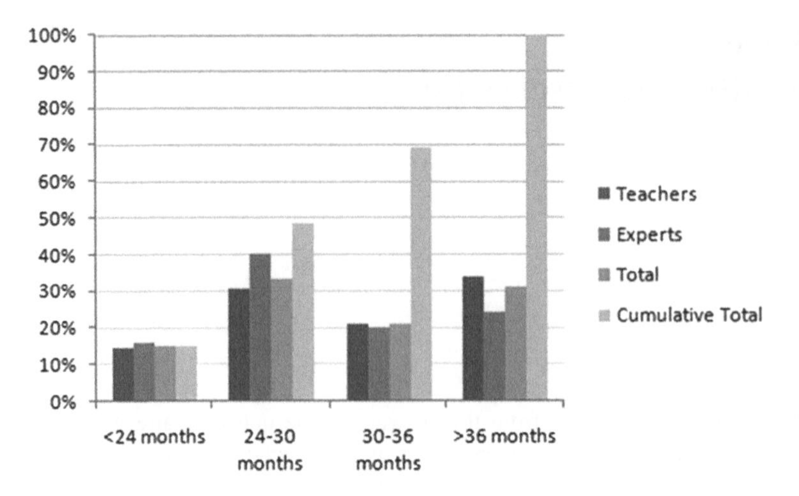

Fig. 4.1 Teachers and experts—understand and follow simple instructions

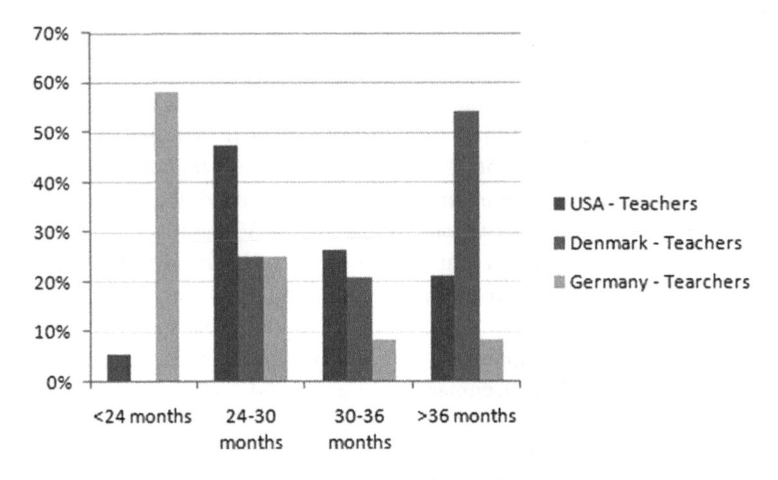

Fig. 4.2 Teachers—understand and follow simple instructions

calculation shows that the majority of children are able to walk down stairs without support or using the hand rail by the age of 30 months, see (Fig. 4.7). The ability of children to walk down the stairs is relevant in case of a fire evacuation for day-care centers located at higher levels of a building (Figs. 4.8 and 4.9).

Regarding children's reaction to unusual events (i.e., not becoming upset), the results from the teachers' answers cannot be generalized, Fig. 4.10. Twenty-nine percentages of the teachers think that children can react without being upset in case of a fire evacuation between 24 and 30 months, 21 % believe between 30–36 months and 21 % believe between 36 and 42 months. The majority of the

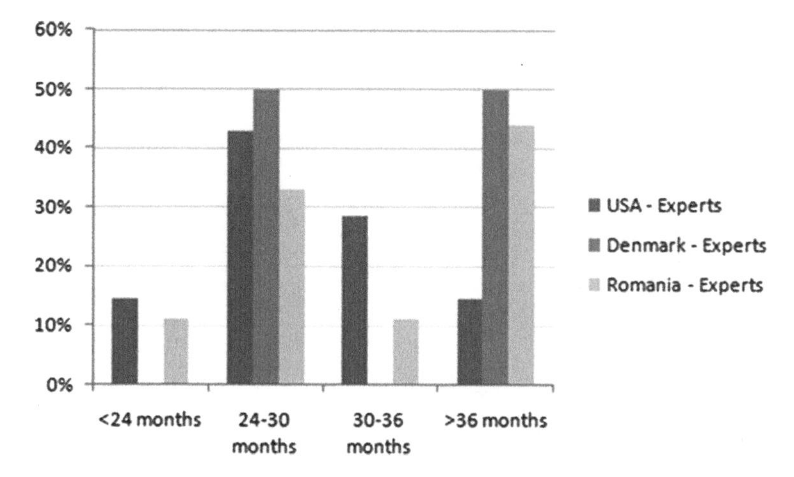

Fig. 4.3 Experts—understand and follow simple instructions

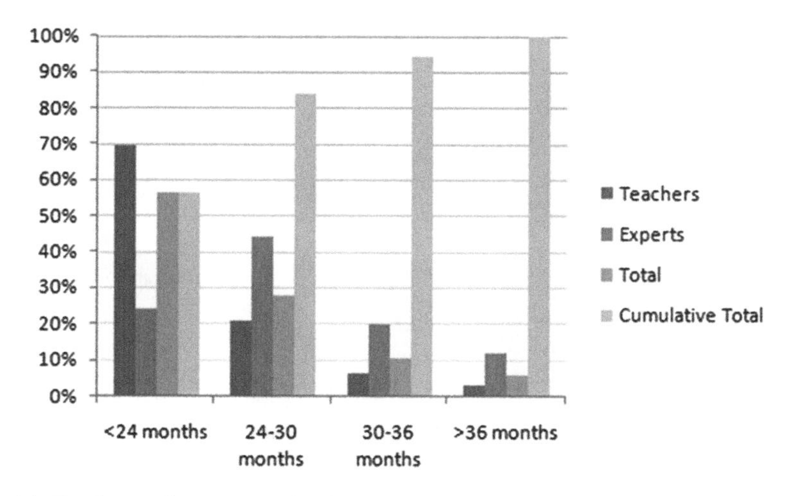

Fig. 4.4 Teachers and experts—walk without any physical support on a horizontal surface

experts believe that by the age of 42 months, children can react without being upset in case of an emergency (Figs. 4.10, 4.11 and 4.12).

One participant commented that children's reactions depend on the adult behavior and state of mind. Many teachers react excitedly during fire drills, causing the children to become upset and excited. When teachers are calm, the children tend to behave calmly as well.

Table 4.1 shows the recommended adult-child ratio. Younger children need more assistance and care during a normal day, but also in case of an emergency. Also this ratio directly affects the evacuation process and time. In the following table, it can be observed the recommended ratios between adults and children in

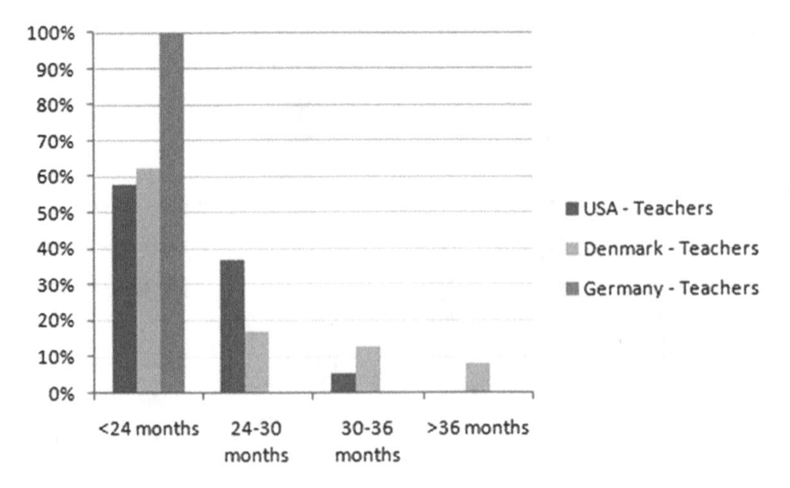

Fig. 4.5 Teachers—walk without any physical support on a horizontal surface

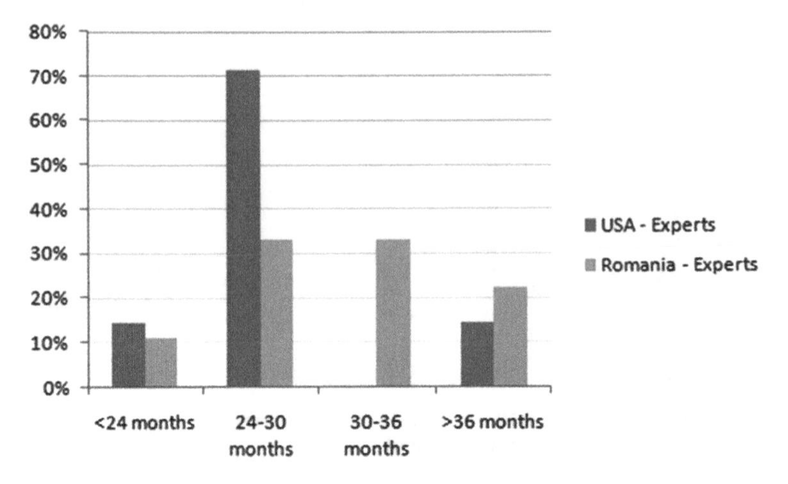

Fig. 4.6 Experts—walk without any physical support on a horizontal surface

day-care centers. To make a comparison, the answers are divided by participant's profession and their country of occupation. The table shows both the mean value and the mode for each category. As can be seen, teachers from Germany suggest a larger ratio for all age intervals, for example: for children under 18 months—1:4 and for children more than 36 months—1:10—using the mean values. The smallest ratios are suggested by experts from US 1:3 for younger children and 1:7 for older children—using the mean values.

An important remark made by one of the participants is that during the day, the normal prescriptive ratio between children and adults is not always met. This happens mostly in the early morning, when part of the staff hasn't arrived yet to

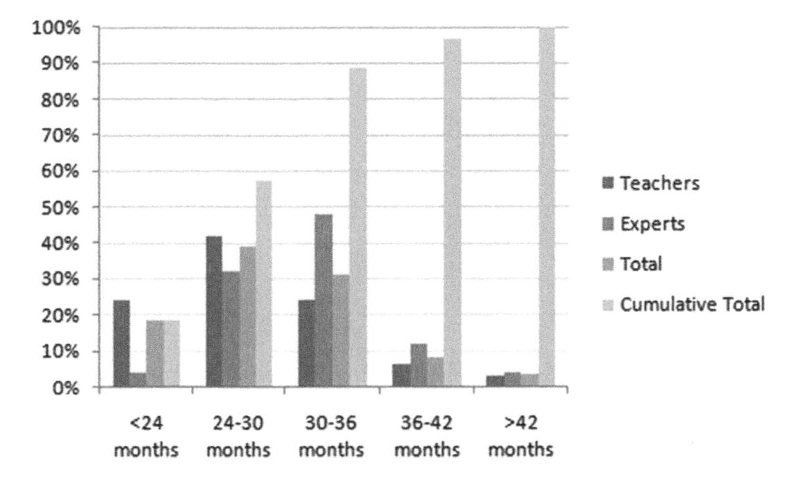

Fig. 4.7 Teachers and experts—walk down the stairs

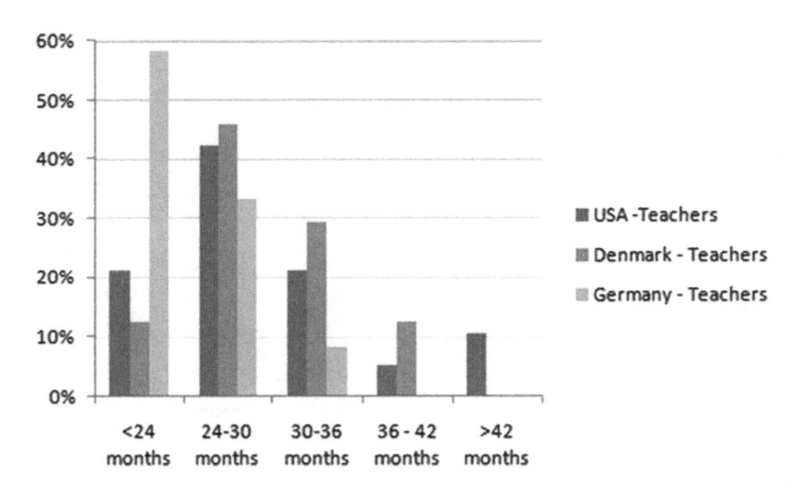

Fig. 4.8 Teachers—walk down the stairs

work or in the late afternoon when some of them have left. In Table 4.2 can be observed that actual ratios in the day-care centers suggested by the German teachers are larger than the ones from US and Denmark for all age intervals.

The largest share of the teachers are able to assist in the evacuation of a facility carrying one child and holding another child's hand at the same time, Fig. 4.13. An observation made by a participant in the survey is that the most common reason that the staff prefers to carry children from younger groups (0–3 years) is because children have a very slow or an insecure walking speed, and in some cases adults prefer to speed up the evacuation.

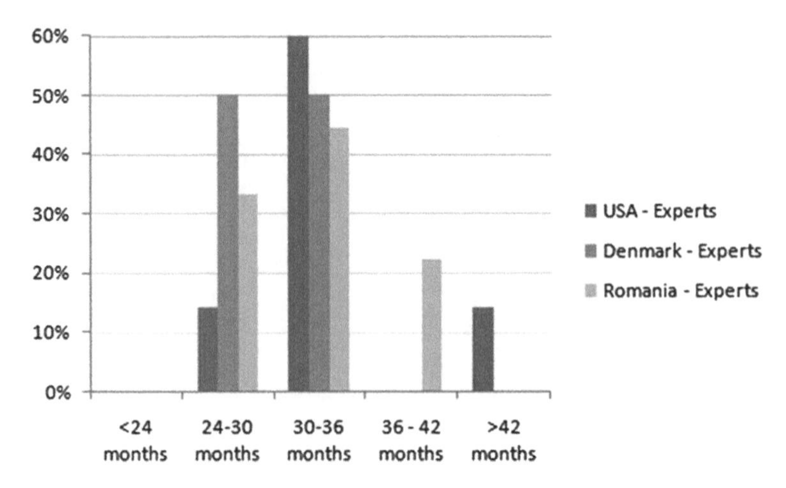

Fig. 4.9 Experts—walk down the stairs

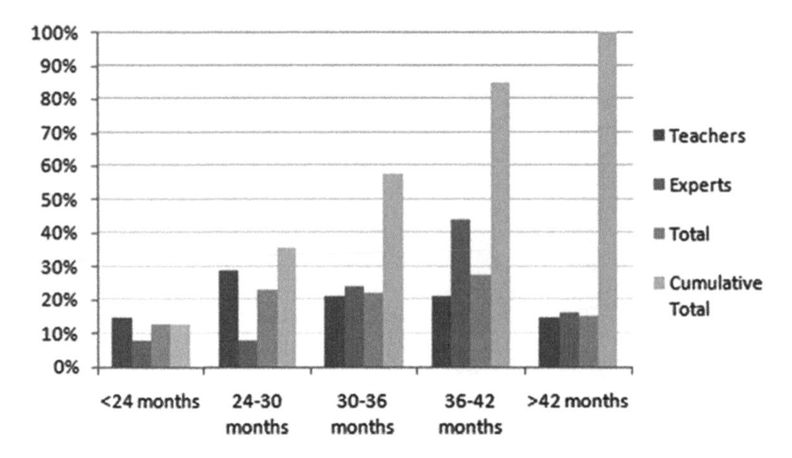

Fig. 4.10 Teachers and experts—children's reaction

Children in day-care centers are used to routines and follow the daily rules. Teachers (45 %) have an optimistic opinion that children react the same when an unfamiliar escape route is used, but experts (67 %) don't have the same opinion, see Fig. 4.14.

Teachers (70 %) and experts (42 %) believe that children can take instruction from other adults whom they are not familiar with, see Fig. 4.15. But in the same time almost as many experts (37 %) do not agree with this statement.

Fifteen percent of the day-care centers represented by the staff who answered the questionnaire do not have a fire alarm system and almost 21 % of them don't have the day-care room located at the ground floor, see Fig. 4.16. These two conditions can result in a dramatically increased total evacuation time.

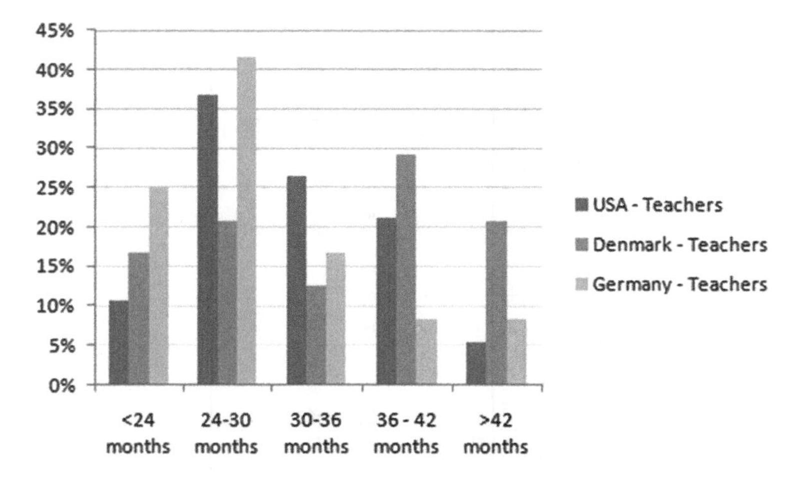

Fig. 4.11 Teachers—children's reaction

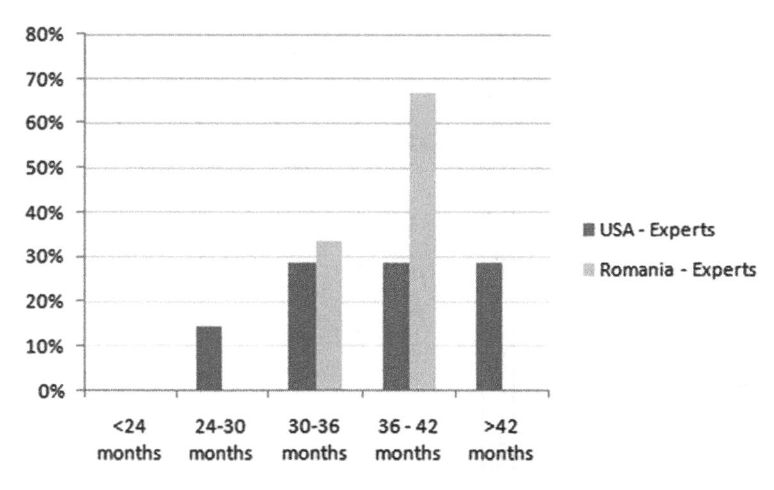

Fig. 4.12 Experts—children's reaction

Five of 16 participants from Denmark, which represents as round 8 % of the total respondents, reported that the institutions where they are working don't have a fire alarm system. This might be because of the fire regulations, which are mandatory only for new buildings. In the same country, 9 of 12 participants answered that the day-care room is not located at the ground floor. Meanwhile in Germany all the answers were positive for the two questions.

Table 4.3 shows that the amount of staff training is larger for USA day-care teachers compared with the European countries. The same observation can be made in regard to fire drills. This can be explained by the different fire regulations.

Table 4.1 Recommended adult-child ratio

Per 1 adult	USA			Denmark			Germany			Romania		
	Teachers [17]		Experts[7]	Teacher [19]		Expert [2]	Teachers [8]		Experts [2]	Experts [9]		
Descriptive Statistics	Mean	Mode	Mean	Mode	Mean	Mode	Mean	Mean	Mode	Mean	Mean	Mode
Number of children under 18 months	3.18	3	3.14	4	3.32	4	2.5	4.25	5	3.5	3.33	2
Number of children between 18 to 24 months	4.41	4	3.43	4	3.95	3	3.5	5.25	5	4	3.67	2
Number of children between 24 to 30 months	5.53	6	4.29	4	4.74	4	3.5	6.25	7	4.5	5.33	4
Number of children between 30 to 36 months	6.41	6	4.71	4	5.68	6	3.5	7.38	10	7	6.44	4
Number of children over 36 months	8.53	8	7.14	10	7.55	8	6	10.13	10	8	8.33	5

Table 4.2 Actual ratios between adults and children

Country	USA		Denmark		Germany	
Descriptive statistics	Mean	Mode	Mean	Mode	Mean	Mode
Number of children under 18 months per 1 adult	3.62	3	3.94	4	6.38	6
Number of children between 18 to 24 months per 1 adult	5.08	4	4.31	4	7.75	6
Number of children between 24 to 30 months per 1 adult	5.80	4	4.94	4	8.50	8
Number of children between 30 to 36 months per 1 adult	7.50	10	6.17	5	9.63	10

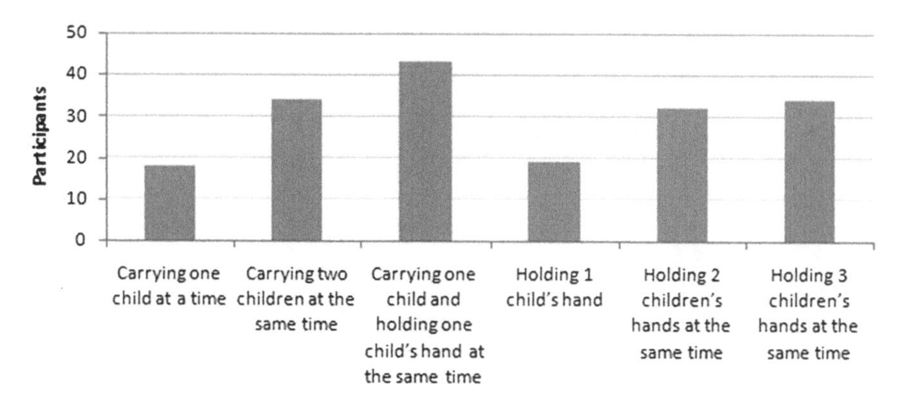

Fig. 4.13 Assisting evacuation

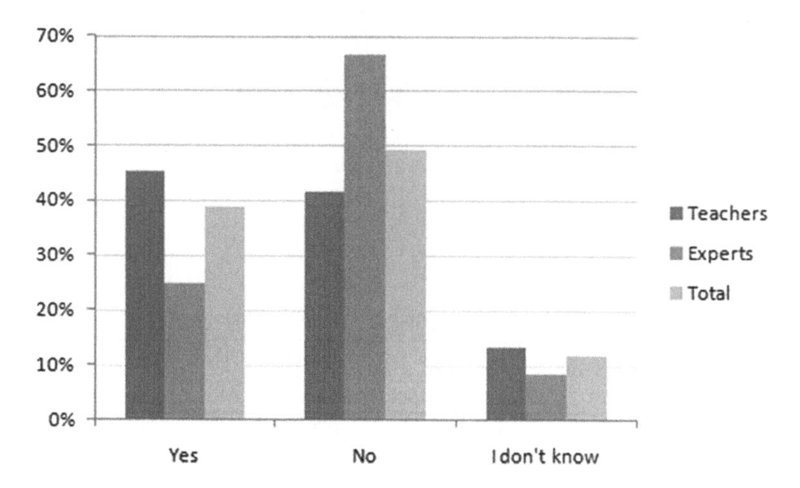

Fig. 4.14 Teachers and experts—unfamiliar escape route

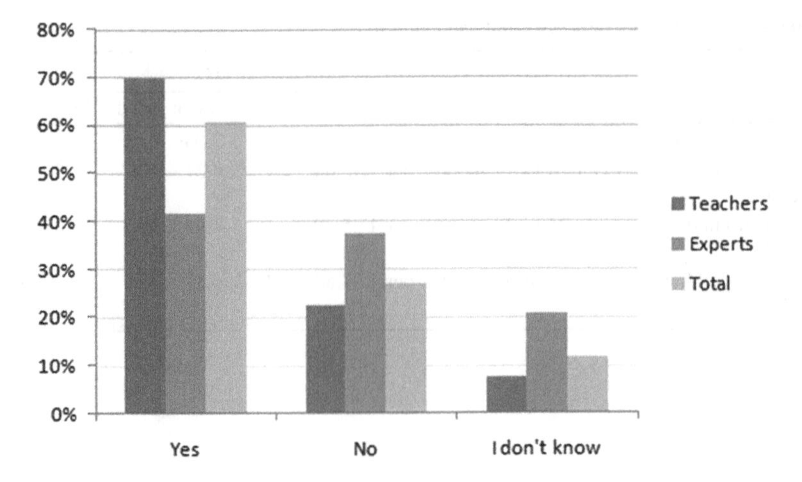

Fig. 4.15 Teachers and experts—instructions

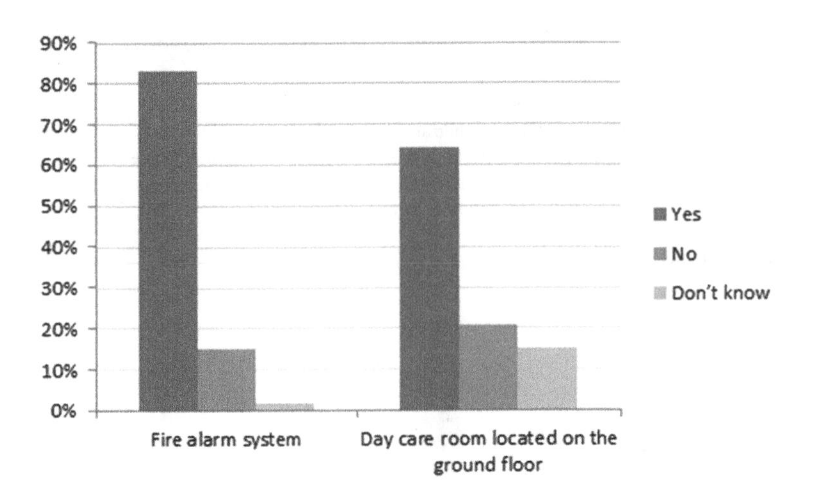

Fig. 4.16 Safety in day-care centers

Table 4.3 Number of fire drills and staff training events in previous year

Country	USA [17]		Denmark [21]		Germany [10]	
Descriptive statistics	Mean	Mode	Mean	Mode	Mean	Mode
Staff training for evacuation	3.35	1	0.71	1	1.3	2
Fire drills	9.06	12	0.95	1	0.5	0

In Denmark, the authorities seem to be afraid to perform fire drills because they might harm and create confusion among children. Fire drills only for the day-care staff are preferred [8].

But even if the code in the United States emphasize that fire drills are mandatory once a month, the mean results shows that not all of the surveyed day-care centers follow the rule.

Chapter 5
Conclusion

The goal of the presented study was primarily to investigate through a questionnaire, distributed in several countries among teachers and early childhood experts, the self-preservation capacity in pre-school children. The second purpose was to create an overview regarding young children behavior in case of an evacuation, adult/children ratio, fire alarm system in day-care institutions, stuff training and fire drills.

A total of 87 participants answered an online survey which collected data for this project, 62 teachers and 25 experts from the USA, Germany, Denmark, Romania, and Canada.

The work has the following findings:

- When between 30 and 36 months, children are considered capable of understanding and following simple instructions.
- When children are less than 24 months old, they are expected to walk on a horizontal plane without assistance.
- When between 24 and 30 months they are expected to walk down the stairs.

Regarding children's behavior in case of an emergency, teachers and experts have differing opinions. The largest share of teachers answered "24–30 months", while the experts answered "36–42 months", to the question concerning the age at which children will not become upset by unusual events. The reason for this difference is the practical versus theoretical experience which the two profession categories have.

When looking at the adult-child ratio, there is a clear difference between children's age groups, with young children needing extra assistance during an evacuation procedure from staff members by carrying or hand holding. This shows the importance of there being enough adults present in order to help children to evacuate.

A big difference between countries was found with respect to the fire alarm systems installed, the training of the staff and number of executed fire drills. A change within the fire regulations and guidelines would be needed in order to change the situation in Europe.

A. Taciuc and A. S. Dederichs, *Determining Self-Preservation Capability in Pre-School Children*, SpringerBriefs in Fire, DOI: 10.1007/978-1-4939-1080-9_5,

Appendix A
Questionnaire Day-Care Center

1. What country do you work in?

2. How many years of experience do you have working in early childhood development?

   ____

3. At what age (in months) can the majority of children understand and follow simple instructions from the staff in case of evacuation? (Ex: "Everyone line up and follow me out!")

   ☐ <24 months
   ☐ 24–30 months
   ☐ 30–36 months
   ☐ >36 months

4. At what age (in months) can the majority of children walk without any physical support on horizontal surface? (Ex: not holding hands or not being carried)

   ☐ <24 months
   ☐ 24–30 months
   ☐ 30–36 months
   ☐ >36 months

5. When are the majority of children able to walk down the stairs alone or using the hand rail? (Ex: not crawling or scooting or sliding up or down stairs)

   ☐ <24 months
   ☐ 24–30 months
   ☐ 30–36 months
   ☐ 36–42 months
   ☐ >42 months

A. Taciuc and A. S. Dederichs, *Determining Self-Preservation Capability in Pre-School Children*, SpringerBriefs in Fire, DOI: 10.1007/978-1-4939-1080-9,

6. At what age do the majority of children react without being upset (Ex: crying, screaming, running) in case something is out of their regular schedule?

   ☐ <24 months
   ☐ 24–30 months
   ☐ 30–36 months
   ☐ 36–42 months
   ☐ >42 months

7. In your professional opinion, what should be the recommended ratio between adults and children? (for all age groups)

   Number of children under 18 months per 1 adult_____
   Number of children between 18 and 24 months per 1 adult_____
   Number of children between 24 and 30 months per 1 adult_____
   Number of children between 30 and 36 months per 1 adult_____
   Number of children over 36 months per 1 adult_____

8. What are the actual ratios between adults and children in your workplace?

   Number of children under 18 months per 1 adult_____
   Number of children between 18 and 24 months per 1 adult_____
   Number of children between 24 and 30 months per 1 adult_____
   Number of children between 30 and 36 months per 1 adult_____
   Number of children over 36 months per 1 adult_____

9. In which of the following ways would you be able to assist in the evacuation of a facility (check as many as apply):

   ☐ Carrying one child at a time
   ☐ Carrying two children at the same time
   ☐ Carrying one child and holding one child's hand at the same time
   ☐ Holding one child's hand
   ☐ Holding two children's hands at the same time
   ☐ Holding three children's hands at the same time

10. In the case that an unfamiliar escape route is used, in your professional opinion, will children react the same as they do to a familiar escape route? (Ex: if primary escape route is blocked or unavailable could the children use fire escape stairs, alternative stair cases, etc.)

   ☐ Yes
   ☐ No
   ☐ I do not know

11. In your professional opinion, would children take instructions from other adults whom they are not familiar with? (Ex: parents of other children, firefighters)?

☐ Yes
☐ No
☐ I do not know

12. At your workplace do you have:

	Yes	No	Don't know
Fire alarm system			
Day-care room located on the ground floor			

13. How many fire drills or staff training for evacuation did you have at your workplace within last year:

Staff training for evacuation	
Fire drills	

Thank you for your assistance with this survey project!

Appendix B
Questionnaire for Children Development Experts

1. What country do you work in?

2. How many years of experience do you have working in early childhood development?

   ____

3. At what age (in months) can the majority of children understand and follow simple instructions from the staff in case of evacuation? (Ex: "Everyone line up and follow me out!")

   ☐ <24 months
   ☐ 24–30 months
   ☐ 30–36 months
   ☐ >36 months

4. At what age (in months) can the majority of children walk without any physical support on horizontal surface? (Ex: not holding hands or not being carried)

   ☐ <24 months
   ☐ 24–30 months
   ☐ 30–36 months
   ☐ >36 months

5. When are the majority of children able to walk down the stairs alone or using the hand rail? (Ex: not crawling or scooting or sliding up or down stairs)

   ☐ <24 months
   ☐ 24–30 months
   ☐ 30–36 months
   ☐ 36–42 months
   ☐ >42 months

6. At what age do the majority of children react without being upset (Ex: crying, screaming, running) in case something is out of their regular schedule?

   ☐ <24 months
   ☐ 24–30 months
   ☐ 30–36 months

A. Taciuc and A. S. Dederichs, *Determining Self-Preservation Capability in Pre-School Children*, SpringerBriefs in Fire, DOI: 10.1007/978-1-4939-1080-9,

☐ 36–42 months
☐ >42 months

7. In your professional opinion, what should be the recommended ratio between adults and children? (for all age groups)

Number of children under 18 months per 1 adult_____
Number of children between 18 and 24 months per 1 adult_____
Number of children between 24 and 30 months per 1 adult_____
Number of children between 30 and 36 months per 1 adult_____
Number of children over 36 months per 1 adult_____

8. In the case that an unfamiliar escape route is used, in your professional opinion, will children react the same as they do to a familiar escape route? (Ex: if primary escape route is blocked or unavailable could the children use fire escape stairs, alternative stair cases, etc.)

☐ Yes
☐ No
☐ I do not know

9. In your professional opinion, would children take instructions from other adults whom they are not familiar with? (Ex: parents of other children, firefighters)?

☐ Yes
☐ No
☐ I do not know

Thank you for your assistance with this survey project!

Appendix C
Email

Dear Participant,

My name is Anca Taciuc and I am an international student at the Technical University of Denmark (DTU). I am working on a project called 'Determining the Self-Preservation Capability in Pre-School Children', in collaboration with the Fire Protection Research Foundation (USA), under the supervision of Associate professor Anne Dederichs at DTU.

The objective of this project is to give guidance to the Technical Committee on Assembly and Educational Occupancies of NFPA's Life Safety Code© to help them determine at what age children are capable of self-preservation.

This will help the Life Safety Code provide requirements for Day-Care Centers that are safer and allow for a more efficient response in the event of a fire.

My task is to survey several Day-Care Center operators and early childhood experts from several countries to determine at what age children would be considerable capable of evacuating themselves.

The survey contains questions regarding the age of children incapable of self-preservation.

Your answers will be treated anonymously and confidentially.

The survey is web based, it takes less than 5 min and can be accessed through this link:

https://www

You are welcome to contact me for further questions or comments.

Thank you for your time and interest!

Sincerely regards!
AncaTaciuc
xxx@student.dtu.dk
Phone: 0045 xxx xxxxxx

A. Taciuc and A. S. Dederichs, *Determining Self-Preservation Capability in Pre-School Children*, SpringerBriefs in Fire, DOI: 10.1007/978-1-4939-1080-9,

Appendix C
Email

Appendix D
Other Tables

The data displayed in the figures in the report are presented in this appendix.

1. At what age (in months) can the majority of children understand and follow simple instructions from the staff in case of evacuation? (Ex: “Everyone line up and follow me out!”)

Table D.1 Teachers and experts

Interval (months)	Teachers	Experts	Total	Cumulative total (%)
<24	15 % (9)	16 % (4)	15 % (13)	15
24–30	31 % (19)	40 % (10)	33 % (29)	48
30–36	*21 %* *(13)*	*20 %* *(5)*	*21 %* *(18)*	*69*
>36	33 % (21)	24 % (6)	31 % (27)	100

Table D.2 Results from different countries

Interval (months)	USA		Denmark		Germany		Romania
	Teachers [19] (%)	Experts [7] (%)	Teachers [24] (%)	Experts [2] (%)	Teachers [12] (%)	Experts [2] (%)	Experts [9] (%)
<24	5	14	0	0	58	0	11
24–30	47	43	25	50	26	100	33
30–36	26	29	20	0	8	0	11
>36	22	14	54	50	8	0	44

A. Taciuc and A. S. Dederichs, *Determining Self-Preservation Capability in Pre-School Children*, SpringerBriefs in Fire, DOI: 10.1007/978-1-4939-1080-9,

2. At what age (in months) can the majority of children walk without any physical support on horizontal surface? (Ex: not holding hands or not being carried)

Table D.3 Teachers and experts

Interval (months)	Teachers	Experts	Total	Cumulative total (%)
<24	*69 %*	*24 %*	*56 %*	*56*
	(43)	*(6)*	*(49)*	
24–30	21 %	44 %	28 %	84
	(13)	(11)	(24)	
30–36	7 %	20 %	10 %	94
	(4)	(5)	(9)	
>36	3 %	12 %	6 %	100.00
	(2)	(3)	(5)	

Table D.4 Results from different countries

Interval (months)	USA		Denmark		Germany		Romania
	Teachers [19] (%)	Experts [7] (%)	Teachers [24] (%)	Experts [2] (%)	Teachers [12] (%)	Experts [2] (%)	Experts [9] (%)
<24	58	14	63	0	100	0	11
24–30	37	72	17	100	0	50	33
30–36	5	0	12	0	0	50	33
>36	0	14	8	0	0	0	22

3. When are the majority of children able to walk down the stairs alone or using the hand rail? (Ex: not crawling or scooting or sliding up or down stairs)

Table D.5 Teachers and experts—walk down stairs

Interval (months)	Teachers	Experts	Total	Cumulative total (%)
<24	24 %	4 %	19 %	18
	(15)	(1)	(16)	
24–30	*42 %*	*32 %*	*39 %*	*57*
	(26)	*(8)*	*(34)*	
30–36	24 %	48 %	31 %	88
	(15)	(12)	(27)	
36–42	7 %	12 %	8 %	96
	(4)	(3)	(7)	
>42	3 %	4 %	3 %	100
	(2)	(1)	(3)	

Table D.6 Results from different countries—walk down stairs

Interval (months)	USA		Denmark		Germany		Romania
	Teachers [19] (%)	Experts [7] (%)	Teachers [24] (%)	Experts [2] (%)	Teachers [12] (%)	Experts [2] (%)	Experts [9] (%)
<24	21	0	11	0	58	0	0
24–30	42	14	49	50	34	50	33
30–36	21	72	29	50	8	50	45
36–42	5	0	11	0	0	0	22
>42	11	14	0	0	0	0	0

4. At what age do the majority of children react without being upset (Ex: crying, screaming, running) in case something is out of their regular schedule?

Table D.7 Teacher and experts—children's reaction

Interval (months)	Teachers	Experts	Total	Cumulative total (%)
<24	15 % (9)	8 % (2)	13 % (11)	13
24–30	28 % (18)	8 % (2)	23 % (20)	36
30–36	*21 %* *(13)*	*24 %* *(6)*	*22 %* *(19)*	*58*
36–42	21 % (13)	44 % (11)	28 % (24)	86
>42	15 % (9)	16 % (4)	14 % (13)	100

Table D.8 Results from different countries—children's reaction

Interval (months)	USA		Denmark		Germany		Romania
	Teachers [19] (%)	Experts [7] (%)	Teachers [24] (%)	Experts [2] (%)	Teachers [12] (%)	Experts [2] (%)	Experts [9] (%)
<24	11	0	17	0	25	0	0
24–30	37	13	21	0	42	50	0
30–36	26	29	13	0	17	0	33
36–42	21	29	29	50	8	0	67
>42	5	29	20	50	8	50	0

Bibliography

NFPA 101, 2012.

Evarts B. (2011) NFPA's "Educational Properties", (http://www.nfpa.org/itemDetail.asp?categoryID=2653&itemID=58156).

Srinivasan G., 87 children die in school fire, The Hindu Online, July 2004, News article (http://www.hindu.com/2004/07/17/stories/2004071707570100.htm).

Xinhua, Death toll from Mexico's daycare center fire rises to 47, People's Daily Online, June 2009, News article (http://english.peopledaily.com.cn/90001/90777/90852/6684347.html).

Murozaki Y. and Ohnishi K. (1985) A study on fire safety and evacuation planning for pre-schools and day care centers, Memoirs of the Faculty of Engineering Kobe University, 32:99–109.

Capote J. et al. (2012) Children Evacuation: Empirical data and egress modelling, Fifth International Symposium on Human Behaviour in Fire.

Larusdottir A.R. and Dederichs A. (2012) Evacuation of children: Movement on Stairs and on Horizontal plane, Fire technology 48(1): 43–53.

Larusddttir A.R. and Dederichs A. (2009) Master Thesis: Evacuation process of daycare centers for children 0–6 years, Technical University of Denamrk, Lyngby.

Building Regulation 10, The Danish Ministry of Economic and Business Affairs, 2010. (http://www.erhvervsstyrelsen.dk/file/155699/BR10_ENGLISH.pdf.).

Kholshchevnikov V.V. et al. (2012) Study of children evacuation from pre-school education institutions, Fire and Materials 36(6–6): 349–366.

Kholshchevnikov V.V. et al. (2009) Pre-school and school children building evacuation, Proceedings of Fourth International Symposium on Human Behaviour in Fire 2009, Robison College, Cambridge, UK.

Campanella M.C. et al. (2011) Empirical data analysis and modelling of the evacuation children from three multi-storey day-care centers, Evacuation and Human Behavior in Emergency Situations EVAC11, Santander, Spai.

Boyse K. and Mohammed L. (2012) Developmental Milestones. Your Child Development & Behaviour Resources, University of Michigan—Heath System (http://www.med.umich.edu/yourchild/topics/devmile.htm).

How Kids Develop. Iowa State University. [Online] December 2006. (https://www.extension.iastate.edu/4hfiles/VI950902FAgesStages.PDF).

Department of Heath and Human Services. Rules for licensing of nursery schools. Child Care Licensing Unit, Maine, USA, 2004.

Canada Goverment of Ontario. Day Nurseries Act, r.r.o. 1990, Regulation 262, Service Ontario E-Laws, 2007.

Metodologie de organizare si functionare a creselor si a altor unitati de educatie timpurie anteprescolara, HG 1252/2012, Constituia Romaniei, Monitorul Oficial, Partea I, nr. 8, Januarie 2013 (In Romanian).

A. Taciuc and A. S. Dederichs, *Determining Self-Preservation Capability in Pre-School Children*, SpringerBriefs in Fire, DOI: 10.1007/978-1-4939-1080-9,

Brown B. et al. (2008) Gender differences in children's pathways to indepedent mobility, London: Children's Geographies 6:1473–3285.

Mackett R. et al. (2007) Children's Independent Movement in the Local Environment, Built Environemnt 33(4): 454–468.

Klupfel H. et al. (2003) Comparison of an Evacuation Exercise in a Primary School to Simulation Results, Traffic and Granular Flow Flow'01, pp 549–554.

CPSIA information can be obtained at www.ICGtesting.com
Printed in the USA
LVOW01s1803260514

387233LV00005B/82/P

9 781493 910793